Ages of stars

STARS—THEIR AGES, LIFESPANS, TIMES TILL

THEIR DEATHS, AND MORE

JOHN HEBERT

3/22/18-5/25/18;

12/26-28/18;

1/3/19

for-meryem, cathy, bunny, spunky, guy, and baby

The universe is very big. If our earth were the size of a smoke particle, the nearest star would be located 2 miles away. The distance spanning across the universe would be equivalent to going the distance of 3.875 million times around the earth. Astronomically, time is very long. If an inch were to equal a year, than a rather short lived star that lives 100 million years would be the distance of 1,578 miles. If

75 years were equal to an inch, such a star's lifespan would be 21 miles long. inch= Our sun's lifespan would be a distance of going 6.31 times around the earth (75 years=1 inch:

100 million years=21 miles), and there are stars whose lifetime distances would be a distance of going 6,310 times around the earth. The number of stars in the universe is very large. If a star were equal in size to a grain of sand, then all the stars in the universe would make a compact ball 12 miles large. Each person on earth receiving an equal number of stars from the total number of them would each have a ball of stars 40 feet in diameter. Stars can be very large. If the earth were an inch large, the largest of star would have diameters 2.8 miles large. About infinite timespan— here is my incomplete attempt to describe infinity using our immortalness as souls. Let us say the each atom were equal to a human lifespan. There are 525 billion atoms which make up a smoke particle. There are also 240 million of these which spans 1 inche's length.

There are 1.6 billion of these inches needed to go around the earth. There are 3.1 billion times *that distance needed to go to the nearest star. Then, there are 21.6 billion of those distances needed to span the distance across the universe. If we were to add up all of the total of our lifetimes (the atom's being a lifetime),our souls and consciousness es will be experiencing life of some kind life during all that time, and that total time is just the beginning and really like being equal to zero time because eternity never ends. We will always be alive and conscious because we are immortal. I wonder what we will be doing during time. I feel it will not be static*

existence but rather one of growth

centering on love.

GIANT STARS
Star's name—
mimosa
beta crucis

*Stellar class—
B.5(3)
color— blue
constellation—
crux
Apparent
magnitude— 1.25*

Absolute magnitude— -2.82
distance 280 light years
mass— 16 suns
radius— 8.4 suns
luminosity— 34,000 suns

Surface temperature— 27,000 degrees kelvin

Velocity of motion — 9.672 miles/second recession

Movement in arcseconds/year—.0162 seconds/year

Direction on the unit circle— 247.07 degrees

parallax—11.71 milliarcseconds

Size in arcseconds — .09836

Gravitational acceleration —201 feet and 8.4 inches/second^2

Escape velocity— 374.13 miles/second

age— 8-11 million years old

lifespan— less than 9 million 760 thousand years

Main sequence (prime of its life) lifespan— less

than 8 million 790 thousand years

Time left as a main sequence star — already completed

When will this star die?— less than 1 million 760 thousand years, *and could supernova at any time*

Fate of the star— supernova, neutron star

Mimosa, a B2(5) star, is 10 solar masses with a

separation distance of 5.4-12 AU from mimosa. Mimosa is losing a sun's mass every 100 million years from strong stellar winds which have a velocity of 1,240 miles/ second.

*Star's name—
bellatrix
gamma orionis*

Stellar class—B2(3)

constellation—orion

color— blue

Right ascension—05 hours 25 minutes 07.86323 seconds

declination— 06 degrees 20 minutes 58.9318 seconds

Apparent magnitude— 1.64

Absolute magnitude— -2.78

distance— 250 light years

mass— 8.6 suns

radius— 20 suns

luminosity— 9,211 suns

Surface temperature — 22,000 degrees kelvin

Velocity of motion — 11.284 miles/second approach

Movement in arcseconds/year—.015 seconds/year

Direction on the unit circle— 212.2 degrees

parallax—43.46 milliarcseconds

size— .2608 arc seconds

gravity— 19 feet 1.2 inches/second^2

Escape velocity— 177.76 miles/second

age— 25 million 200 thousand years old

lifespan—less than 27 million 720 thousand years

Main sequence (prime of its life) lifespan— less

than 24 million 948 thousand years

Time left as a main sequence star — star has already completed this phase of its lifetime

When will this star die?— less than 2

million 520
thousand years

Fate of the star—
supernova, white
dwarf star

This star is much
closer to us than
the 3 main belt
stars in the orion

the hunter constellation.

*Star's name—
alcyone
eta tauri*

Stellar class—B5(3)

color— blue

constellation—taurus

Apparent magnitude— 2.87

Absolute magnitude— -2.62
distance— 440 light years
mass— 3.4-3.8 suns
radius— 8.2 suns
luminosity— 2,030 suns

Surface temperature — 12,258 degrees kelvin

Velocity of motion — 3.348 miles/second approach

Movement in arcseconds/year—.0437 seconds/year

Direction on the unit circle— 286.54 degrees

parallax—8.09 milliarcseconds

Size in arcseconds — .06634

Gravitational acceleration — 47 feet 6.95 inches/second^2

Escape velocity — 179.62 miles/second

Rotation velocity— 92.38 miles/second

age— cluster is located in is 100 million years old

lifespan— less than 355 million 260 thousand-469

million 140 thousand years

Main sequence (prime of its life) lifespan— less than 319 million 730 thousand-422 million 230 thousand years

Time left as a main sequence star — should already left the main sequence, but 100 million year age implies still a main sequence star.

When will this star die?— less than 35 million 530 thousand-46 million 910 thousand years, but age of 100 million years gives time of death at 255 million 260 thousand-369 million 140 thousand years from now

Fate of the star— planetary nebula, while dwarf

*Star's name—
kaus australis
epsilon sagittarii*

Stellar class— B9.5(3)

color— blue

constellation— sagittarius

Apparent magnitude— 1.95

Absolute magnitude— -1.41
distance— 143 light years
mass— 3.515 suns
radius— 6.8 suns
luminosity— 363 suns

Surface temperature—9,960 degrees kelvin

Velocity of motion — 9.3 miles/second approach

Movement in arcseconds/year—.13 seconds/year

Direction on the unit circle— 99.57 degrees

parallax—22.76 milliarcseconds

Size in arcseconds — .1548

Gravitational acceleration — 67 feet 6.65 inches/second^2

Escape velocity — 194.9 miles/second

Rotation velocity— 146.32 miles/second

age— 232 million years old

lifespan—less than 431 million 700 thousand years

Main sequence (prime of its life) lifespan— less than 156 million 530 thousand years

Time left as a main sequence star — past the main

sequence, but age too young, so should still be a main sequence star

When will this star die?— less than 199 million 700 thousand years

Fate of the star— planetary nebula, white dwarf

Star has companion— mass .95 suns, absolute magnitude 4.96, apparent magnitude 8.17, luminosity .89 suns, separation distance 155 AU (14.415 billion miles), period of orbit 1,929.73 years

*Star's name—
miaplacidus
beta carinae*

*Stellar class—
A1(3)*

color— blue white

constellation— carina

Apparent magnitude— 1.69

Absolute magnitude— -1.03

distance— 113.2 light years

mass— 3.5 suns

radius— 6.8 suns

luminosity— 288 suns

Surface temperature — 8,866 degrees kelvin

Velocity of motion — 3.224 miles/second approach

Movement in arcseconds/year— .11 seconds/year

Direction on the unit circle— 141.28 degrees

parallax—28.82 milliarcseconds

Size in arcseconds — .19598

Gravitational acceleration — 67 feet 3.24 inches/second^2

Escape velocity — 194.48 miles/second

age— 260 million years old

lifespan—less than 2 billion 693 million years

Main sequence (prime of its life) lifespan— less

than 2 billion 424 million years

Time left as a main sequence star — less than 2 billion 164 million years

When will this star die?— less than 2

billion 433 million years

Fate of the star— planetary nebula, white dwarf star

Star's name—
cursa

beta eridani

*Stellar class—
A3(5)*

color— blue white

*constellation—
eridanus*

Apparent magnitude— 2.796

Absolute magnitude— .59

distance— 81 light years

mass— 2.0 suns
radius— 2.4 suns
luminosity— 25 suns

Surface temperature—

8,360 degrees kelvin

Velocity of motion — 3.6 miles/second approach

Movement in arcseconds/year—.0757 seconds/year

Direction on the unit circle— 222.99 degrees

parallax—36.5 milliarcseconds

Size in arcseconds — .0876

Gravitational acceleration —308 feet and 4.2 inches/ second^2

Escape velocity— 398.89 miles/ second

*Rotation velocity—
121.52 miles/
second*

*age— less than 1
billion 591 million
years old*

*lifespan—less than
1 billion 768 million
years*

Main sequence (prime of its life) lifespan— less than 1 billion 591 million years

Time left as a main sequence star — already left the main sequence

When will this star die?— less than 176 million 800 thousand years

Fate of the star— planetary nebula, white dwarf

*Star's name—
rasalhahue
alpha ophiucus*

Stellar class— A5(3)

color— blue white

constellation— ophiucus

Apparent magnitude— 2.07

Absolute magnitude— 1.248

distance— 48.6 light years

mass— 2.4 suns

radius— 2.6 suns

luminosity— 25.1-25.6 suns

Surface temperature— 7,880-8,050 degrees kelvin

Velocity of motion — 7.812 miles/second recession

Movement in arcseconds/year— .2264 seconds/year

Direction on the unit circle— 288.89 degrees

parallax—67.13 milliarcseconds

Size in arcseconds — .067

Gravitational acceleration —315 feet and 9 inches/second^2

Escape velocity— 260.45 miles/second

*Rotation velocity—
148.8 miles/
second*

*age—740-800
million years old*

*lifespan—less than
1 billion 121 million
years*

Main sequence (prime of its life) lifespan— less than 1 billion 9 million years

Time left as a main sequence star — past the main sequence

When will this star die?— less than 321 million-381 million years

Fate of the star— planetary nebula, white dwarf star

This star has a K5-7(5) companion

with mass .85 suns, radius less than 1.5 suns, temperature 4,161 kelvin, luminosity of .566 suns, apparent magnitude of 5.45, absolute magnitude of 6.32, period of orbit of 8.62 years, and separation distance of 390.98 million miles.

*Star's name—
seginus
gamma bootes*

Stellar class— A7(3)

color— blue white

constellation— bootes

Apparent magnitude— 3.3

Absolute magnitude— .91

distance— 86.3 light years

mass— 1.95 suns
radius— 7.46 suns
luminosity— 34 suns

Surface temperature — 7,800 degrees kelvin

Velocity of motion — 22.11 miles/second approach

Movement in arcseconds/year —

.1522 seconds/year

Direction on the unit circle— 121.59 degrees

parallax—37.58 milliarcseconds

Size in arcseconds — .28

Gravitational acceleration — 31 feet 1.66 inches/ second^2

Escape velocity — 138.6 miles/ second

Rotation velocity — 86.8 miles/ second

age— less than 1 billion 695 million years old

lifespan—less than 1 billion 883 million years

Main sequence (prime of its life) lifespan— less

than 1 billion 695 million years

Time left as a main sequence star — past the main sequence

When will this star die? — less than

188 million 327 thousand years

Fate of the star—planetary nebula, white dwarf

*Star's name—
adhafara
zeta leonis*

*Stellar class—
F0(3)
color— white
constellation— leo
Apparent magnitude— 3.33
Absolute magnitude— -1.19*

distance— 274 light years

mass— 3.0 suns
radius— 6.0 suns
luminosity— 85 suns

Surface temperature— 6,792 degrees kelvin

Velocity of motion — 9,673 miles/second approach

Movement in arcseconds/year—

.00684 seconds/year

Direction on the unit circle— 337.33 degrees

parallax—11.9 milliarcseconds

Size in arcseconds — .0714

Gravitational acceleration — 74 feet 9.144 millimeters/second^2

Escape velocity — 191.68 miles/second

age— less than 444 million 750 thousand years old

lifespan— less than 494 million 200 thousand years

Main sequence (prime of its life)

lifespan— less than 444 million 750 thousand years

Time left as a main sequence star — past the main sequence

When will this star die?— less than 49 million 450 thousand years

Fate of the star— planetary nebula, white dwarf star

*Star's name—
caph
beta cassiopeiae*

*Stellar class—
F2(3)*

color— yellow white

*constellation—
Cassiopeia*

Apparent magnitude— 2.28

Absolute magnitude— 1.3
distance— 54.7 light years
mass— 1.91 suns
radius— 3.43-3.69 suns
luminosity— 27.3 suns

Surface temperature— 7,079 degrees kelvin

Velocity of motion — 7.006 miles/second

Movement in arcseconds/year— 2.97 seconds/year

Direction on the unit circle— 338.94 degrees

parallax—59.58 milliarcseconds

Size in arcseconds — .212

Gravitational acceleration — 134 feet and 1.2 inches/ second^2

Escape velocity— 196.56 miles/ second

*Rotation rate—
1.12 times daily*

age— 1 billion 90 million-1 billion 180 million years old

lifespan—less than 1 billion 983 million years

Main sequence (prime of its life) lifespan— less than 1 billion 785 million years

Time left as a main sequence star — less

than 605 million-695 million years

When will this star die?— less than 803 million-893 million years

*Fate of the star—
planetary nebula,
white dwarf*

*Star's name—
sadalbari
mu pegasi*

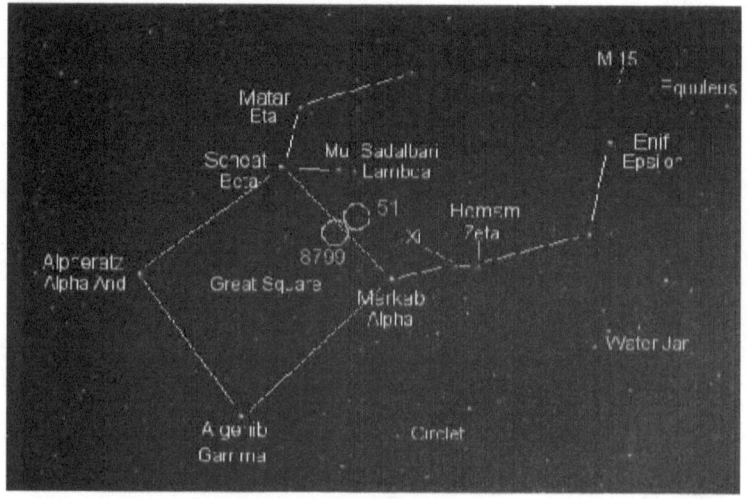

*Stellar class—
G8(3)*

color— yellow

*constellation—
pegasus*

*Apparent
magnitude— 3.514*

*Absolute
magnitude— .432*

distance— 106.1 light years

mass— 1.3 suns
radius— 9.6 suns
luminosity— 57.45 suns

Surface temperature—

4,950 degrees kelvin

Velocity of motion — 8.395 miles/second recession

Movement in arcseconds/year — .1506seconds/year

Direction on the unit circle— 342.07 degrees

parallax—30.74 milliarcseconds

Size in arcseconds — .2951

Gravitational acceleration — 12

feet 6.43 inches/ second^2

Escape velocity— 160.9 miles/ second

age— less than 4 billion 671 million years old

lifespan— less than 5 billion 190 million years

Main sequence (prime of its life) lifespan— less than 4 billion 671 million years

Time left as a main sequence star — already left main sequence

When will this star die? — 519 million years

*Fate of the star—
planetary nebula,
white dwarf star*

Star's name— vindamiatrix epsilon virginis

Stellar class— G8(3)

color— yellow

constellation— virgo

Apparent magnitude— 2.826

Absolute magnitude— .37

distance— 109.6 light years

mass— 2.64 suns
radius— 10.6 suns
luminosity— 77 suns

Surface temperature—

5,086 degrees kelvin

Velocity of motion — 9.052 miles/second approach

Movement in arcseconds/year — .0207 seconds/year

Direction on the unit circle— 177.37 degrees

parallax—29.76 milliarcseconds

Size in arcseconds — .3155

Gravitational acceleration — 20

feet 10.56 inches/second^2

Escape velocity— 135.28 miles/second

age— 470-650 million years old

lifespan— less than 880 million

600 thousand years

Main sequence (prime of its life) lifespan— less than 794 million 750 thousand years

Time left as a main sequence star — already left the main sequence

When will this star die? — less than 230 million 600

thousand–410 million 600 thousand years

Fate of the star— planetary nebula, white dwarf star

Star's name—
menkent
theta centuari

Stellar class—
K0(3)

color— orange

constellation— centaurus

Apparent magnitude— 2.06

Absolute magnitude— .87

distance— 58.8 light years

mass— 1.57 suns
radius— 10.6 suns
luminosity— 60 suns

Surface temperature— 4,980 degrees kelvin

Velocity of motion — .806 miles/second recession

Movement in arcseconds/year— .58398 seconds/year

Direction on the unit circle — 220.16 degrees

parallax — 55.45 milliarcseconds

Size in arcseconds — .58778

Gravitational acceleration — 12

feet 5 inches/
second^2

Escape velocity—
104.33 miles/
second

age— less than 2
billion 911 million
years old

lifespan—less than 3 billion 238 million years

Main sequence (prime of its life) lifespan— less than 2 billion 914 million years

Time left as a main sequence star — already finished main sequence

When will this star die? — less than 328 million 800 thousand years

Fate of the star—planetary nebula, white dwarf star

Soft x-rays emitted by this star estimate energy 1.4×10^{27} ergs/second

*Star's name—
deneb kaitos
beta ceti*

*Stellar class—
K0(3) color—
orange*

constellation— cetus

Apparent magnitude— 2.02

Absolute magnitude— -.13

distance— 96.3 light years

mass— 2.8 suns radius— 16.78 suns luminosity— 139.1 suns

Surface temperature— 4,797 degrees kelvin

Velocity of motion — 8.998 miles/second recession

Movement in arcseconds/year — .0328 seconds/year

Direction on the unit circle — 9.7 degrees

parallax — 33.86 milliarcseconds

Size in arcseconds — .0568

Gravitational acceleration — 8

feet 10 inches/ second^2

Escape velocity— 188.6 miles/ second

age— less than 1 billion years old

lifespan— less than 763 million

260 thousand years

Main sequence (prime of its life) lifespan— less than 686 million years

Time left as a main sequence star —

already left main sequence

When will this star die?— less than 76 million 225 thousand years

Fate of the star— planetary nebula, white dwarf star

Star evolved away from being an A type main sequence star and has beed fusing helium in its core for over 100 million years.

Star's name—
arcturus
alpha bootes

Stellar class—
K0(3)

color— orange

constellation— bootes

Apparent magnitude— -.05

Absolute magnitude— -.3

distance— 36.7 light years

*mass— 1.08 suns
radius— 25.4 suns
luminosity— 170 suns*

Surface temperature— 4,286 degrees kelvin

Velocity of motion — 3.22 miles/second approach

Movement in arcseconds/year — 2.96 seconds/year

Direction on the unit circle — 201.86 degrees

parallax—88.83 milliarcseconds

Size in arcseconds — 2.256

Gravitational acceleration — 1 feet 5.85 inches/second^2

Escape velocity— 90.16 miles/ second

age— 5.9-8.6 (6.9 billion) billion years old

lifespan— less than 8 billion 250 million years

Main sequence (prime of its life) lifespan— less than 7 billion 425 million years

Time left as a main sequence star — already left main sequence

When will this star die?— less than 825 million years

Fate of the star— planetary nebula, white dwarf star

This star will make its closest approach to us in

4000 years and will be a few hundredths of a light years closer to us than it is today.

Star's name— capella

alpha aurigae

Stellar class— K0(3)/ G1(3)

color— orange/yellow

constellation— auriga

Apparent magnitude— .08/-.43

Absolute magnitude—

.296/.167

distance— 42.919 light years

mass— 2.5687/2.4828 suns

radius— 11.98/8.83 suns

luminosity— 5.6/73.33 suns

Surface temperature— 4,970 degrees kelvin

Velocity of motion — 18.563 miles/second recession

Movement in arcseconds/year— .428 seconds/year

Direction on the unit circle— 271.14 degrees

parallax—76.3 milliarcseconds

Size in arcseconds — .9141

Gravitational acceleration — 15 feet 10.86 inches/second^2

Escape velocity — 202.46 miles/second

age— 590-650 million years old

lifespan— less than 945 million 600 thousand years

Main sequence (prime of its life) lifespan— less

than 851 million years

Time left as a main sequence star — already left main sequence

When will this star die? — less than

94 million 600 thousand years

Fate of the star— planetary nebula, white dwarf star

The capellian star system is one of the brightest x-ray sources in the sky.

*Star's name—
pollux
beta geminorum*

Stellar class— K0(3)

color— orange

constellation— gemini

Apparent magnitude— 1.14

Absolute magnitude— 1.18

distance— 33.76 light years

mass— 1.91 suns
radius— 8.8 suns
luminosity— 43 suns

Surface temperature — 4,666 degrees kelvin

Velocity of motion — 2 miles/ second recession

Movement in arcseconds/year— .054 seconds/year

Direction on the unit circle— 265.36 degrees

parallax—96.54 milliarcseconds

Size in arcseconds — .84955

Gravitational acceleration — 21 foot 11 inches/second^2

Escape velocity — 96.54 miles/second

*Rotation rate—
once every 558
days*

*age— 724 million
years old*

*lifespan— less
than 1 billion 983
million years*

Main sequence (prime of its life) lifespan— less than 1 billion 785 million years

Time left as a main sequence star — already left main sequence

When will this star die?— less than 1 billion 259 years

Fate of the star— planetary nebula, white dwarf star

Has an extrasolar planet (thestias) with a mass of 2.3

jupiters, a period of 589.64 days, separation distance of 2.05 AU (190.756 million miles), and an eccentricity of .02.

*Star's name—
Ankaa
alpha phoenicis*

*Stellar class—
K.5(3)*

color— orange

constellation— phoenicis

Apparent magnitude— 2.377

Absolute magnitude— .52

distance— 84.8 light years

mass— estimate 4.88 suns

radius— 15 suns

luminosity— 52.98 suns

Surface temperature— 4,436 degrees kelvin

Velocity of motion — 46.25 miles/second recession

Movement in arcseconds/year — .366 seconds/year

Direction on the unit circle — 296.88 degrees

parallax—38.5 milliarcseconds

Size in arcseconds — .5775

Gravitational acceleration — 4 feet 5.23 inches/second^2

*Escape velocity—
74.17 miles/
second*

*age— less than
171 million 246
thousand years old*

*lifespan—less than
190 million 274
thousand years*

Main sequence (prime of its life) lifespan— less than 171 million 246 thousand years

Time left as a main sequence star — already left main sequence

When will this star die? — less than 19 million 27 thousand years

Fate of the star— planetary nebula, white dwarf

*Star's name—
hamel
alpha Arietis*

Stellar class—K1(3)

color— orange

constellation— aries

Apparent magnitude— 2.0

Absolute magnitude— .47

distance— 65.8 light years

mass— 1.5 suns

radius— 14.9 suns

luminosity— 55.47 suns

Surface temperature— 4,480 degrees kelvin

Velocity of motion — 8.804 miles/second approach

Movement in arcseconds/year—

.1507 seconds/year

Direction on the unit circle— 317.62 degrees

parallax—49.56 milliarcseconds

Size in arcseconds — .7384

Gravitational acceleration — 6 feet/ second^2

Escape velocity— 138.73 miles/ second

age— 1.5-5.3 billion years old

lifespan—less than 3 billion 629 million years

Main sequence (prime of its life) lifespan— less than 3 billion 266 million years

Time left as a main sequence star — already left main sequence

When will this star die? — less than 263 million years

Fate of the star— planetary nebula, white dwarf

Extrasolar planet Hamel b

Mass >= 1.8 jupiters, period 380.8 days, separation

distance 1.0645 AU (99 million miles), eccentricity .25.

*Star's name—
errai

gamma cephei*

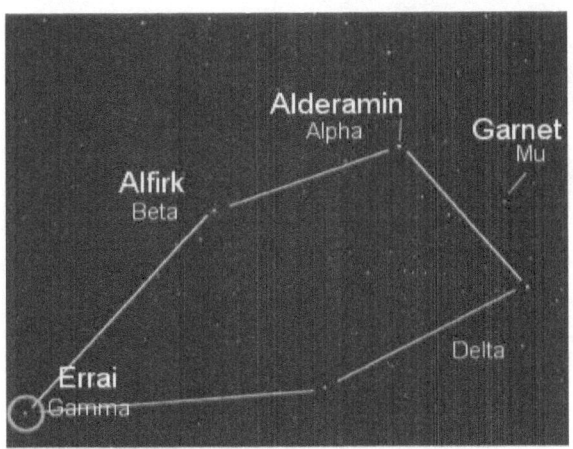

Stellar class—k1(3/4)

color—orange

constellation—cepheus

Apparent magnitude— 3.22

Absolute magnitude— 2.62
distance— 44.9 light years
mass— 1.4 suns
radius— 4.79 suns
luminosity— 7.66 suns

Surface temperature— 4,883 degrees kelvin

Velocity of motion — 5.456 miles/second approach

Movement in arcseconds/year—

.1267 seconds/year

Direction on the unit circle — 103.06 degrees

parallax — 72.69 milliarcseconds

Size in arcseconds — .3482

Gravitational acceleration — 54 feet 2.4 inches/second^2

Escape velocity — 146.55 miles/second

age— 2 billion 620 million-3 billion 880 million years old

lifespan—less than 4 billion 312 million years

Main sequence (prime of its life) lifespan— less

than 3 billion 881 million years

Time left as a main sequence star — already left main sequence (could have

810,000-1,261,000 years left on main sequence)

When will this star die?— less than 432 million-1 billion 692 million years

Fate of the star—planetary nebula, white dwarf

This star has a M4(5) red dwarf companion gamma cephei b with apparent magnitude 6.2,

5.51 absolute magnitude, .54 sun's luminosity, mass .409, period of orbit 67.5 years, separation distance 1.499 billion miles. Extrasolar planet gamma cephei ab (tadmor) is orbiting the primary star. It has a mass of 1.85 jupiters, a period of orbit of 903.3 days, an orbital distance of 169.053 million miles, and an eccentricity of .049. The primary star will become the north pole star 3000-4000 AD. In 5200 AD, iota cephei will be the north pole star.

Star's name— avior epsilon carinae

Stellar class— K3(3)/ B2(5)

color— orange/ blue

constellation— carina

Apparent magnitude— 2.166/4.1

Absolute magnitude— -4.47/-2.26

distance— 610 light years

mass— 9/7.3 suns

radius— 201.85/2.17 suns

luminosity— 5,250.28/686.06 suns

Surface temperature— 3,523/20,417 degrees kelvin

Velocity of motion — 1.192 miles/second recession

Movement in arcseconds/year—.02273 seconds/year

Direction on the unit circle — 133.69 degrees

parallax — 5.39 milliarcseconds

Size in arcseconds — 1.088

Gravitational acceleration — 2

feet 9 inches/second^2

Escape velocity— 92.33 miles/second

age— 21.1-41.3 million years old

lifespan—less than 41 million 150 thousand years

Main sequence (prime of its life) lifespan— less than 37 million years

Time left as a main sequence star — already left main sequence

When will this star die? — less than 4 million 150 thousand years

Fate of the star—supernova, white dwarf

The period of this binary system is 785 days and has a separation distance of 4 AU (372 million miles).

*Star's name—
kochab*

beta ursae minoris

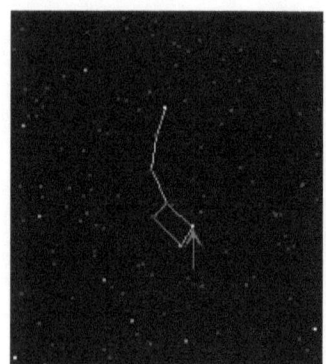 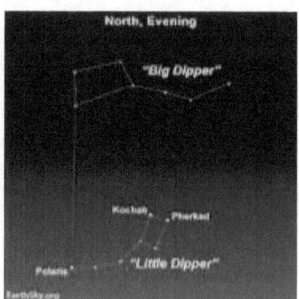

*Stellar class—
K4(3)*

color— orange

constellation— ursa minor

Apparent magnitude— 2.08

Absolute magnitude— -.83

distance— 130.9 light years

mass— 2.2 suns

radius— 42.06 suns

luminosity— 390 suns

Surface temperature— 4,030 degrees kelvin

Velocity of motion — 10.515 miles/second recession

Movement in arcseconds/year — .0405 seconds/year

Direction on the unit circle — 93.51 degrees

parallax — 24.91 milliarcseconds

Size in arcseconds — 1.0477

Gravitational acceleration — 1

feet 1.26 inches/ second^2

Escape velocity— 99.996 miles/ second

age— less than 1 billion 154 million years old

lifespan—less than 1 billion 393 million years

Main sequence (prime of its life) lifespan— less than 1 billion 154 million years

Time left as a main sequence star — already left main sequence

When will this star die? — less than 239 million years

Fate of the star—planetary nebula, white dwarf

Extrasolar planet beta ursa minoris b—

*Mass 6.1 jupiters
Period 522.3 days*

Separation distance 1.71 AU (159.03 million miles)

Eccentricity .19

*Star's name—
aldebaran*

alpha tauri

Stellar class— K5(3)

color— orange constellation— taurus

Apparent magnitude—

.86

Absolute magnitude— -.641

distance— 68.3 light years

mass— 1.5 suns

radius— 44.13 suns

luminosity— 518 suns

Surface temperature— 3,910 degrees kelvin

Velocity of motion — 33.64 miles/second recession

Movement in arcseconds/year— .19 seconds/year

Direction on the unit circle— 280.53 degrees

parallax—49.97 milliarcseconds

Size in arcseconds — 2.2052

Gravitational acceleration — 8 inches 5.43 millimeters/second^2

Escape velocity—80.61 miles/second

age— less than 3 billion 266 million years old

lifespan—less than 3 billion 629 million years

Main sequence (prime of its life) lifespan— less than 3 billion 266 million years

Time left as a main sequence star — already left main sequence

When will this star die?— less than 363 million years

Fate of the star— planetary nebula, white dwarf

The star is losing 1 earth mass every 300,000 years.

Earth seen from aldebaran would be magnitude 6.4 between constellations ophiucus and scorpius. Pioneer 10 is going in the direction of aldebaran and will make its closest approach in 2 million years.

*Star's name—
mirach
beta andromedae*

Stellar class—M0(3)

color— red

constellation— Andromeda

Apparent magnitude— 2.05

Absolute magnitude— -1.76

distance— 197 light years

mass— 3-4 suns

radius— 100 suns

luminosity— 1,995 suns

Surface temperature—

3,842 degrees kelvin

Velocity of motion — .0372 miles/second

Movement in arcseconds/year— .1139 seconds/year

Direction on the unit circle — 323.85 degrees

parallax — 16.52 milliarcseconds
Size in arcseconds — 1.652

Gravitational acceleration — 3

inches 18.6 millimeters/second^2

Escape velocity— 81.8 miles/second

age— less than 392 million 700 thousand years old

lifespan—less than 436 million 345 thousand years

Main sequence (prime of its life) lifespan— less than 392 million 700 thousand years

Time left as a main sequence star — already left main sequence

When will this star die? — less than 43 million 645 thousand years

*Fate of the star—
planetary nebula,
white dwarf*

Star's name — yed prior

delta ophiucus

Stellar class— M.5(3)

color— red

constellation— ophiucus

Apparent magnitude— 2.75

Absolute magnitude—-.9

distance— 171 light years

mass— 1.5 suns
radius— 59 suns
luminosity—195.93 suns

Surface temperature— 3,679 degrees kelvin

Velocity of motion — 1.2338 miles/second approach

Movement in arcseconds/year— .1429 seconds/year

Direction on the unit circle— 190.47 degrees

parallax—19.06 milliarcseconds

Size in arcseconds — 1.1245

Gravitational acceleration — 4 inches 15.12 millimeters/second^2

*Escape velocity—
69.72 miles/
second*

*age— less than 3
billion 266 million
years old*

*lifespan— less
than 3 billion 629
million years*

Main sequence (prime of its life) lifespan— less than 3 billion 266 million years

Time left as a main sequence star — already completed main sequence

When will this star die?— 363 million years

Fate of the star— planetary nebula, white dwarf

Star's name— menkar alpha ceti

Stellar class— M1.5(3)

color— red

constellation— cetus

Apparent magnitude— 2.53

Absolute magnitude —-3.2

distance— 249 light years

mass— 2.3 suns
radius— 89 suns
luminosity— 1,455 suns

Surface temperature— 3,795 degrees kelvin

Velocity of motion — 16.1696 miles/second approach

Movement in arcseconds/year — .07685 seconds/year

Direction on the unit circle— 178.57 degrees

parallax—13.09 milliarcseconds

Size in arcseconds — 1.165

Gravitational acceleration — 3

inches 2.45 millimeters/ second^2

Escape velocity— 70.3 miles/ second

age— less than 1 billion 122 million years old

lifespan— less than 1 billion 247 million years

Main sequence (prime of its life) lifespan— less than 1 billion 122 million years

Time left as a main sequence star — already completed main sequence

When will this star die? — less than 124 million 700 thousand years

*Fate of the star—
planetary nebula,
white dwarf*

Star's name—
propus
eta geminorum

*Stellar class—
M2(3)/G*

*color— red
constellation—
gemini*

*Apparent
magnitude—
3.15/6*

Absolute magnitude— -3.9/.67

distance— 380 light years

mass— 6.3 (originally an M2-8(5) star)/ 2.06/1.18 suns

radius— 153 suns

luminosity— 3,162/46.26 suns

Surface temperature— 3,548 degrees kelvin

Velocity of motion —no information found

Movement in arcseconds/year— .01214 seconds/year

Direction on the unit circle — 257.8 degrees

parallax — 8.48 milliarcseconds

Size in arcseconds — 1.297

Gravitational acceleration — 2

inches 22.1 millimeters/second^2

Escape velocity— 88.72 miles/second

age— less than 511 million 425 thousand years old

lifespan— less than 567 million 50 thousand years

Main sequence (prime of its life) lifespan—

less than 511 million 425 thousand years old

Time left as a main sequence star — already completed main sequence

When will this star die?— less than 56 million 750 thousand years

Fate of the star—planetary nebula, white dwarf

One set of stars—Period 53.23 years Distance 1.316 billion miles Other set of stars—Period 474 years

Distance 5.65 billion miles

Spectroscopic binary— Period 2,983 days

Distance 371.916 million miles

Star's name—
gacrux
gamma crucis

Stellar class—M3.5(3)

color— red

constellation— crux

Apparent magnitude— 1.64

Absolute magnitude— -.52

distance— 88.6 light years

mass— 1.5 suns
radius— 84 suns
luminosity— 820 suns

Surface temperature— 3,626 degrees kelvin

Velocity of motion — 12.77 miles/second recession

Movement in arcseconds/year — .265 seconds/year

Direction on the unit circle — 266.78 degrees

parallax — 36.83 milliarcseconds

Size in arcseconds — 3.094

Gravitational acceleration — 2 inches 6.78 millimeters/second^2

*Escape velocity—
58.43 miles/
second*

*age— less than 3
billion 266 million
years old*

*lifespan— less
than 3 billion 629
million years*

Main sequence (prime of its life) lifespan— less than 3 billion 266 million years old

Time left as a main sequence star — already completed main sequence

When will this star die?— less than 363 million years

Fate of the star— planetary nebula, white dwarf

This star is a red giant branch rather than an asymptotic

giant branch red giant star.

*Star's name—
mira
omicron ceti*

Stellar class— M7(3)

color— red

constellation— cetus

Apparent magnitude— 6.57 (2-10.1)

Absolute magnitude— .99 (-2.82 to 5.28)

distance— 300 light years

mass— 1.18 suns

radius— 332 suns

luminosity— .66-1,148.55 suns

Surface temperature— 2,918 degrees kelvin

Velocity of motion — 39.56 miles/second recession

Movement in arcseconds/year— .2374 seconds/year

Direction on the unit circle— 97.5 degrees

parallax—10.87 milliarcseconds

Size in arcseconds — 3.622

Gravitational acceleration — 2.9 millimeters/second^2

Escape velocity — 16.16 miles/second

age— 6 billion years old

lifespan— less than 6 billion 611 million years

Main sequence (prime of its life) lifespan— less

than 5 billion 950 million years

Time left as a main sequence star — already completed main sequence (left the main sequence 50 million years ago.)

When will this star die?— 661 million years

Fate of the star— planetary nebula, white dwarf

Binary star system with a red giant variable, Mira a,

and a white dwarf, mira b. They are separated by 70 AU (6.61 billion miles) and have a period of 585.66 years. Mira a is a pulsating variable *star, the first non supernova pulsating star discovered, with the possible exception of Algol. The period between pulses is 10,000 years, and when each pulse's luminosity increases, the pulse grows stronger. Mira shed its 13 light year long tail 10s of thousands of years ago.*

The star is moving 130 kilometers/ second (80.83 miles/ second). Mira b could be a k type main sequence star of mass .7 suns.

BRIGHT GIANT STARS

Star's name—
mintaka
delta orionis

*Stellar class—
O9.5(2)/ B1(5)/
B0(4)/B3(5)*

*constellation—
orion*

color— blue

*Right ascension—
05 hours 32*

minutes 00.40009 seconds

declination— 00 degrees 17 minutes 56.7242 seconds

Apparent magnitude— 2.23 (2.5/3.9)/ 4.93/3.63

Absolute magnitude— -5.8/-5.4/-2.9/-4.2

distance— 1,200 light years

mass— 24/8.4/22.5/9 suns

*radius—
16.5/6.5/10.4/5.7
suns*

*luminosity—
190,000/16,000/
63,000/ 3,300 suns*

Surface temperature — 29,500 degrees kelvin

Velocity of motion — 11.47 miles/second recession

Movement in arcseconds/year —

.00094 seconds/year

Direction on the unit circle— 313.85 degrees

parallax—43.46 milliarcseconds

size— .0448 arc seconds

gravity— 78 feet 4.1 inches/second^2

Escape velocity— 326.94 miles/second

age— less than 3 million 544 thousand years old

lifespan—less than 3 million 898 thousand/ 48 million 899 thousand/28 million 669 thousand/ 128 million 918 thousand years

Main sequence (prime of its life) lifespan— less than 3 million 544 thousand/44 million 10 thousand/ 25 million 800 thousand/116

million 30 thousand years

Time left as a main sequence star — star has already completed this phase of its lifetime

When will this star die? — less than

354 thousand/4 million 890 thousand/2 million 867 thousand/ 12 million 892 thousand years

Fate of the star— supernova, black hole/ planetary

nebula, white dwarf/supernova, neutron star/planetary nebula, white dwarf

B1(5) orbit 5.73 days period and 5.83 million miles distance

B0(4) orbit 4.84 hours period and 65,689 miles distance (results ??)

*Star's name—
canopus
alpha carinae*

*Stellar class—
A9(2)*

(yellow bright giant star)

constellation— canis major

color— blue white

Right ascension— 06 hours 23 minutes 57.10988 seconds

declination— -52 degrees 41 minutes 44.3810 seconds

Apparent magnitude— -.74

Absolute magnitude— -5.71

distance — 310 light years

mass — 8 suns

radius — 71 suns

luminosity — 10,700 suns

Surface temperature — 6,998 degrees kelvin

Velocity of motion — 7.8 miles/second recession

Movement in arcseconds/year— .233 seconds/year

Direction on the unit circle— 49.38 degrees

parallax—10.55 milliarcseconds

size— .747 arc seconds

gravity— 1 foot 10.59 inches/second^2

Escape velocity— 90.99 miles/second

age— less than 55 million 240 thousand old

lifespan—less than 60 million 764 thousand years

Main sequence (prime of its life)

lifespan— less than 55 million 240 thousand years

Time left as a main sequence star — star has already completed this phase of its lifetime

When will this star die?— less than 5 million 524 thousand years

Fate of the star— planetary nebula, white dwarf canopus' orbital distance extends is comparable to a distance all the way to the planet mercury. There are no other stars closer than canopus more luminous, and it has been the brightest star during 3 epochs over the past 4 million years. 9000 years ago, Sirius became the brightest star and will remain so for another 210,000 years. In 480,000 years, will again become the brightest star, and will remain so for 510,000 years.

*Star's name—
adhara
epsilon canis majoris*

Stellar class—B2(2)

constellation—canis major

color— blue

Right ascension—06 hours 58 minutes 37.6 seconds

declination— -28 degrees 58 minutes 19 seconds

Apparent magnitude— 1.5/7.5

Absolute magnitude— -4.8/1.8

distance— 430.5 light years

mass— 12.6 suns

radius— 13.9 suns

luminosity— 38,700/154.8 suns

Surface temperature — 22,900 degrees kelvin

Velocity of motion — 16 miles/second recession

Movement in arcseconds/year —

.00349 seconds/year

Direction on the unit circle— 41.05 degrees

parallax—7.57 milliarcseconds

size— .1054 arc seconds

gravity— 57 feet 11.55 inches/second^2

Escape velocity— 258.09 miles/second

age— 19 million 900 thousand

years-25 million 100 thousand old

lifespan — less than 21 million 890 thousand years-27 million 610 thousand years

Main sequence (prime of its life)

lifespan— less than 19 million 900 thousand years-25 million 100 thousand years

Time left as a main sequence star — star has already

completed this phase of its lifetime

When will this star die?— less than 1 million 990 thousand years-2 million 510 thousand years

Fate of the star—supernova, neutron star

Adhara's companion star is separated by 5.28 billion miles and has an orbital period of 427.79 years. 4.7 million years ago, it was the brightest star in the sky at magnitude -3.99. no other star in the sky will exceed -3.99 magnitude for at least 5 million years. 4.7 million years ago, it was 34 light years away. This star is the brightest known ultraviolet source in the night sky.

Star's name—
canopus
alpha carinae

Stellar class — A9(2)

(yellow bright giant star)

constellation — canis major

color— blue white

Right ascension—

06 hours 23 minutes 57.10988 seconds declination— -52 degrees 41 minutes

44.3810 seconds

Apparent magnitude—-.74

Absolute magnitude— -5.71

distance— 310 light years

mass— 8
suns radius—
71 suns
luminosity—
10,700 suns
Surface
temperature

— 6,998 degrees kelvin

Velocity of motion — 7.8 miles/ second recession

Movement in arcseconds/year— .233 seconds/year

Direction on the unit circle

— 49.38 degrees parallax — 10.55 milliarc seconds

size— .747 arc seconds

gravity— 1 foot 10.59 inches/second^2

Escape velocity— 90.99 miles/second

age— less than 55

million 240 thousand old lifespan—less than 60 million 764 thousand years

Main sequence (prime of its life) lifespan— less than 55 million 240 thousand years Time left as a main sequence star — star has already

completed this phase of its lifetime

When will this star die? — less than 5 million 524

thousand years
Fate of the star—
planetary nebula, white dwarf

canopus' orbital distance extends is comparable to a distance all the way to the planet mercury. There are no other stars closer than canopus more luminous, and it has been the brightest star during 3 epochs over the past 4 million years.

9000 years ago, Sirius became the brightest star and will remain so for another 210,000 years. In 480,000 years, will again become the brightest star, and will remain so for 510,000 years.

*Star's name—
beta draconis
rastaban*

*Stellar class—
G2(2)*

*constellation—
draco*

color— yellow

*Right ascension—
17 hours 30
minutes 25.96170
seconds*

*declination— 50
degrees 18*

minutes 04.9993 seconds

Apparent magnitude— 2.79

Absolute magnitude— -2.28

distance— 380 light years

*mass— 6 suns
radius— 40 suns
luminosity— 1,000 suns*

Surface temperature— 5,160 degrees kelvin

Velocity of motion — 12.4 miles/second approach

Movement in arcseconds/year — .02 seconds/year

Direction on the unit circle — 142.3 degrees

parallax—8.58 milliarcseconds

size— .343 arc seconds gravity—3 feet 4 inches/second^2

Escape velocity—104.99 miles/second

age— 65 million years old

lifespan—less than 71 million 500 thousand years

Main sequence (prime of its life) lifespan— less

than 65 million thousand years

Time left as a main sequence star — star has already completed this phase of its lifetime

When will this star die?— less than 6

million 500 thousand years

Fate of the star—planetary nebula, white dwarf

This star has a companion white dwarf star. The separation distance is 238.43 AU (22.17 billion miles) and has an orbital period of 4,000 years.

Star's name— omicron Scorpii

Stellar class— A4(2/3)

constellation— scorpius

color— blue white

Right ascension— 16 hours 20

minutes 38.18068 seconds

declination— -24 degrees 10 minutes 09.5491 seconds

Apparent magnitude— 4.74

Absolute magnitude— -4.0
distance— 900 light years
mass— 7.9 suns
radius— 15 suns
luminosity— 3,405 suns

Surface temperature — 8,128 degrees kelvin

Velocity of motion — 5.084 miles/second approach

Movement in arcseconds/year— .0148 seconds/year

Direction on the unit circle— 196.98 degrees

parallax—3.71 milliarcseconds

size— .543 arc seconds gravity— 31 feet 2.43 inches/second^2

Escape velocity— 196.73 miles/second

age— 34 million 900 thousand years-44 million 700 thousand years old

lifespan—less than than 38 million 390 thousand years-49

million 170 thousand years

Main sequence (prime of its life) lifespan— less than 34 million 900 thousand years-44 million 700 thousand years

Time left as a main sequence star — star has already completed this phase of its lifetime

When will this star die? — less than 3 million 490 thousand years-4 *million 470 thousand years*

Fate of the star—planetary nebula, white dwarf

*Star's name—
beta
canis majoris*

Stellar class— B8(2/3)

constellation— canis major

color— blue

Right ascension— 07 hours 03 minutes 45.49305 seconds

declination— -15 degrees 37 minutes 59.83 seconds

Apparent magnitude— 4.1

Absolute magnitude— -1.3

distance— *440 light years*

mass— *3.74 suns*
radius— *5.6 suns*
luminosity—
310.54 suns

Surface temperature— 13,596 degrees kelvin

Velocity of motion — 19.84 miles/ second recession

Movement in arcseconds/year— .01136 seconds/year

Direction on the unit circle— 170.79 degrees

parallax—7.38 milliarcseconds

size—

.0415 arc seconds
gravity— 105 feet 11.8 inches/second^2

Escape velocity— 221.53 miles/second

age— less than 11 million 700-13 million 100 thousand years

lifespan—less than 406 million 640 thousand years

Main sequence (prime of its life)

lifespan— less than 369 million 680 thousand years

Time left as a main sequence star — star has already completed this phase of its lifetime

(this star should still be a main sequence star.)

When will this star die?— less than 393 million 540 thousand-394 million 940 thousand years

Fate of the star— planetary nebula, white dwarf

*Star's name—
sargas
theta scorpii*

*Stellar class—
F0(2)*

*constellation—
scorpius*

*color— yellow
white*

*Right ascension—
17 hours 37*

minutes 19.12985 seconds

declination— -42 degrees 59 minutes 52.1808 seconds

Apparent magnitude— 1.84/5.36

Absolute magnitude— -2.71/.588

distance— 300 light years

mass— 5.66 suns

radius— 26/49.76 suns

luminosity— 1,834 suns

Surface temperature— 7,268 degrees kelvin

Velocity of motion — .868 miles/ second recession

Movement in arcseconds/year—.00636 seconds/year

Direction on the unit circle— 330.61 degrees

parallax—10.86 milliarcseconds

size— .2825 arc seconds

gravity— 7 feet 5.24 inches/second^2

Escape velocity— 126.48 miles/second

age— less than 1 billion 960 million years

lifespan—less than 2 billion 180 million years

Main sequence (prime of its life) lifespan— less

than 1 billion 960 million years

Time left as a main sequence star — star has already completed this phase of its lifetime

When will this star die? — less than 220 million years

Fate of the star — planetary nebula, white dwarf

This star has an oblate shape due to its very rapid

spin. Its equator is 19% lager than at its poles. Theta Scorpio b, its companion, is separated from the primary star by 6.47 arc seconds, or 6.68 million *miles, and has an orbital period of 7 days 21 hours 59 minutes.*

*Star's name—
albireo
beta cygni*

Stellar class—
K2(2)/B8/ B8(5)

constellation—
cygnus

color— orange

Right ascension—
19 hours 30
minutes 43.286
seconds

declination— 27 degrees 57 minutes 34.84 seconds

Apparent magnitude— 3.18/5,82/5.11

Absolute magnitude— -2.45/-.25/-.49

distance— 430 light years

mass— 14.52/3.84/3.7 suns

radius—69/2.49/2.59 suns

luminosity—816.85/107.67/134.46 suns

Surface temperature—4,270/12,000/13,200 degrees kelvin

Velocity of motion — 14.92 miles/second approach

Movement in arcseconds/year — .00945 seconds/year

Direction on the unit circle— 224.87 degrees

parallax—7.51 milliarcseconds

size— .523 arc seconds

gravity— 2 feet 8.523 inches/second^2

Escape velocity— 124.35 miles/second

age— less than 554 million 539 thousand years

(3rd star — 100 million years old)

lifespan—less than 609 million 993 thousand years

Main sequence (prime of its life) lifespan— less than 554 million

539 thousand years

Time left as a main sequence star — star has already completed this phase of its lifetime

When will this star die?— less than

55 million 454 thousand years

Fate of the star— supernova, neutron star

One binary separation distance 24.89 million miles and

period 50.568 days, and 1st and 3rd star separation distance 3 billion 326 million miles with period of orbit 213.859 years.

*Star's name—
matar eta pegasi*

*Stellar class—
G2(2)/ F0(5)*

*constellation—
pegasus*

color— yellow

*Right ascension—
22 hours 43
minutes 00.13743
seconds*

declination—30 degrees 13 minutes 16.4822 seconds

Apparent magnitude— 2.95

Absolute magnitude— -1.18

distance— 167 light years

mass— 3.82 suns

radius— 18 suns

luminosity— 247 suns

Surface temperature— 5,450 degrees kelvin

Velocity of motion — 2.67 miles/ second recession

Movement in arcseconds/year— .01554 seconds/year

Direction on the unit circle— 170.24 degrees

parallax—19.51 milliarcseconds

size— .2687 arc seconds

gravity— 10 feet 5.73 inches/second^2

Escape velocity— 124.88 miles/second

age— 240-300 million years old

lifespan—less than 385 million 686 thousand years

Main sequence (prime of its life) lifespan— less than 350 million

624 thousand years

Time left as a main sequence star — star has already completed this phase of its lifetime

When will this star die? — less than

85 million 686 thousand-145 million 686 thousand years

Fate of the star— planetary nebula, white dwarf

Binary star system with a period of

orbit 813 days and separation distance 7.91 AU, or 735.9 million miles. Eccentricity .183.

Star's name—
scheat

beta pegasi

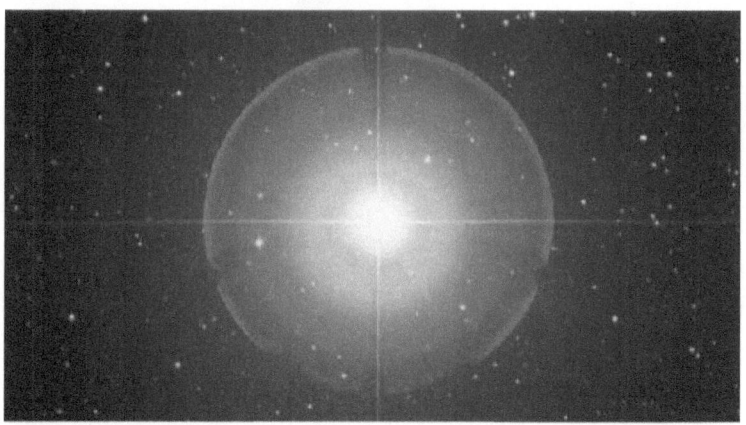

Stellar class—
M2.5(2/3)

*constellation—
pegasus*

color— red

*Right ascension—
23 hours 03
minutes 46.45746
seconds*

*declination—28
degrees 04*

minutes 58.0336 seconds

Apparent magnitude— 2.43

Absolute magnitude— -1.47

distance— 196 light years

mass— 2.1 suns

radius— 95 suns

luminosity— 329.76 suns

Surface temperature— 3,689 degrees kelvin

Velocity of motion — 5.394 miles/second recession

Movement in arcseconds/year — .0194 seconds/year

Direction on the unit circle— 3.97 degrees

parallax—16.64 milliarcseconds

size— 1.58 arc seconds

gravity— 2.48 inches/ second^2

Escape velocity— 40.3 miles/ second

age— less than 1 billion 565 million years

lifespan—less than 1 billion 721 million years

Main sequence (prime of its life) lifespan— less than 1 billion 565 million years

Time left as a main sequence star — star has already

completed this phase of its lifetime

When will this star die?— less than 156 million years

Fate of the star— planetary nebula, white dwarf

A semi regular variable star with a period of 43.3 days. This star is losing 10^-8 solar masses per year and has an expanding shell of mass and dust with a radius of 16 times the sun's radius (13.824 million miles)

SUPERGIANT STARS Stars

star's name—
RW cephei

Stellar class— K2(0-1a)

location— constellation of Cepheus

color— orange

Right ascension— 22 hours 23 minutes 07.01657 seconds

declination— 55

degrees 57 minutes 47.6862 seconds

Apparent magnitude— 6.0

Absolute magnitude— -8.0

distance—11,410 light years

mass— 13.9 suns

radius— 1,535 suns

luminosity— 550,000 suns

Surface temperature— 4,015 degrees kelvin

Movement in arcseconds/year— .00382 arcseconds/ year

Direction on the unit circle— 225.85 degrees

parallax— .81 milliarcseconds/year

size—.439 arc seconds

gravity— 1.6 millimeters/second^2

*Escape velocity—
25.8 miles/ second*

*age— 18 million
700 thousand
years*

*lifespan— less
than 20 million 570
thousand years*

Main sequence (prime of its life) lifespan— less than 18 million 700 thousand years

Time left as a main sequence star — has completed this phase of its life

When will this star die?— less than 1 million 870 thousand years

Fate of the star— supernova, neutron star

Semi regular variable star with a

period of 346 days with magnitude fluctuation of 6 to 7.3. its temperature is intermediate between a red super giant and a *yellow hyper giant star.*

Star's name— HR5171

Stellar class— K0(0-1a)/ B0(1b) (note-there are 2 stars)

location— constellation of Centaurus

colors— orange/ blue

Right ascension— 13 hours 47 minutes 10.875 seconds

declination— -62 degrees 35 minutes 23.06 seconds

Apparent magnitude— 6.1 to 7.5/9.13

Absolute magnitude— -9.2/-5.8

distance—11,700 light years

mass— 27-36/2-20 suns

radius— 1,575/650 suns

luminosity— 630,000/206,000 suns

Surface temperature— 5,000/26,000 degrees kelvin

Velocity through space— 23.684 miles/ second approach

Movement in arcseconds/year— .003885 arcseconds/ year

Direction on the unit circle— 229.18 degrees

parallax— 1.35 milliarcseconds

size— .4388/.18 arc seconds

gravity— (with mass=20)— 12.7 millimeters/second^2

Escape velocity— 38.34/47.55 miles/second

age— less than 1 million 157 thousand-2 million 376 thousand/3 million 500 thousand years

lifespan— less than 1 million 286 thousand-2 million

640 thousand/ 3 million 850 thousand years

Main sequence (prime of its life) lifespan— less than 1 million 157 thousand-2 million 376 thousand years/3 million 500 thousand years

Time left as a main sequence star — has completed this phase of its life

When will this star die?— less than

129 thousand-264 thousand/ 385 thousand years

Fate of the star— supernova, black hole/ supernova, neutron star

This is one of the largest stars

known. It is a contact binary with an orbital period of 1.304 days and a separation distance of 2.256 million miles. It has an initial mass of 32-40 suns

star's name— pz Cassiopeia

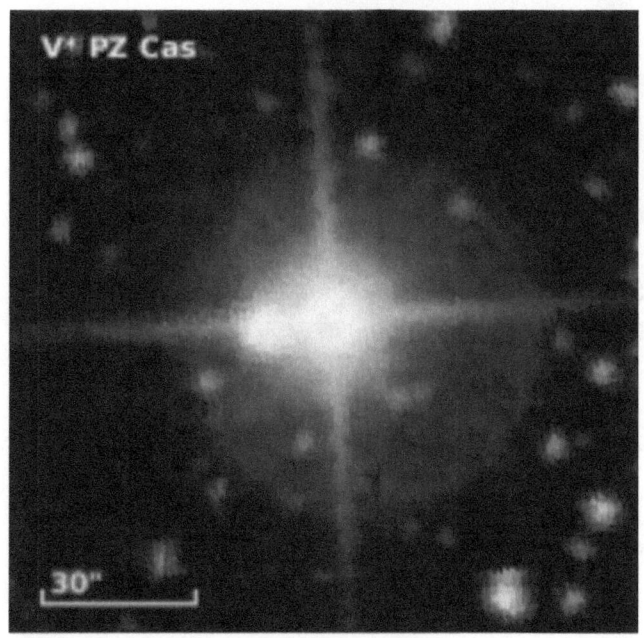

PZ Cassiopeiae

PZ Cassiopeiae is the bright star towards upper right in this WISE infrared image.

Stellar class— M3L

*location—
constellation of
Cassiopeiae*

color— red

*Right ascension—
23 hours 44
minutes 03.28104
seconds*

declination— 61 degrees 47 minutes 22.1823 seconds

Apparent magnitude— 18.9

Absolute magnitude— -7.89

distance—9,160.6 light years

mass— estimate 48.12 suns

radius— 1,190 or 1,940 suns

luminosity— 240,000-270,000 suns

Surface temperature— 3,600 degrees kelvin

Velocity through space— 28.32 miles/ second approach

Movement in arcseconds/year— .00546 arcseconds/ year

Direction on the unit circle— 229.46 degrees

parallax— .356 milliarcseconds

size—

.4235-.69 arc seconds

gravity— 9.2 millimeters/second^2

Escape velocity— 54.51 miles/second

age— 8-10 million years

lifespan— less than 8 million 800 thousand years-11 million years

Main sequence (prime of its life) lifespan— less

than 8-10 million years

Time left as a main sequence star — has completed this phase of its life

When will this star die?— less than 800,000-

1 million years

Fate of the star—supernova, neutron star This is a semi regular variable star.

Star's name— mu cephei

garnet star

Stellar class— M2(1a)

*location—
constellation of
Cepheus*

color— red

*Right ascension—
21 hours 43
minutes 30.4809
seconds*

declination— 56 degrees 46 minutes 48.166 seconds

Apparent magnitude— 4.08

Absolute magnitude— -7.63

distance — 5,000 light years

mass — 19.2 suns

radius — 1,260 suns

luminosity — 283,000 suns

Surface temperature— 3,750 degrees kelvin

Velocity through space— no information found

Movement in arcseconds/year— .00598 arcseconds/ year

Direction on the unit circle— 331.21 degrees

parallax— .55 milliarcseconds

size— .82152 arc seconds

gravity— 3.28 millimeters/second^2

Escape velocity— 33.46 miles/second

age— 9 million 100 thousand-10 million 100 thousand years

lifespan— less than 10 million 100 thousand years-11 million 100 thousand years

Main sequence (prime of its life) lifespan— less than 9 million 100 thousand years-10 million 100 thousand years

Time left as a main sequence star —

has completed this phase of its life

When will this star die?— less than 1 million-1 million 10 thousand years

Fate of the star— supernova, black hole

This star will evolve back to a blue supergiant star, then to a luminous blue variable or wolf rayet star before its core collapse. Its post red *supergiant phase will be a type 2n/ 2b supernova or wolf rayet type 1b/ 1c supernova. It is one of the most luminous red supergiant stars in the milky way and one of the largest to be discovered. A spherical shell of ejected matter extends out to 6 arc seconds, or 1.8968 billion miles with an expansion velocity of 6.2* miles/ second, and is 2000-3000 years old. For an age of 2000-3000 years old and an expansion rate of 6.2 miles/ second, the shell would be 391.3-652.19

billion miles in diameter.

Star's name— IRC-10414

IRC-10414

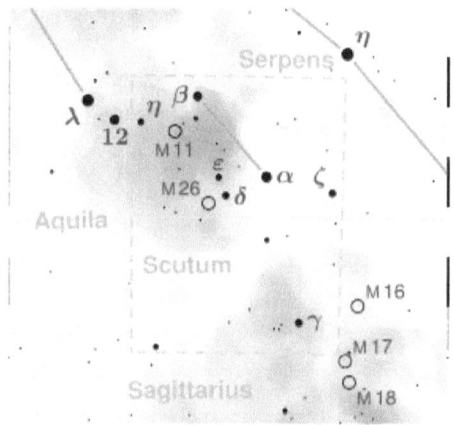

Location of IRC-10414 (and WR 114 (A Wolf-Rayet star)) in

the constellation Scutum.

*Stellar class—
M7L*

*location—
constellation of
Scutum*

color— red

*Right ascension—
18 hours 23*

minutes 17.90 seconds

declination— -13 degrees 42 minutes 47.3 seconds

Apparent magnitude— 12.0

Absolute magnitude— -8.18

distance—3,520 light years

mass— estimate 14.82 suns (initial mass 20-25 suns. Could be up to but less than 47.55 sun masses.)

radius— 1,200 suns

luminosity— 160,000 suns

Surface temperature— 3,300 degrees kelvin

Velocity through space— 17.732 miles/ second

Movement in arcseconds/year—.00582 arcseconds/ year

Direction on the unit circle— 16.5 degrees

parallax— .5 milliarcseconds

size— 1.11 arc seconds

gravity— 2.79 millimeter/second^2

Escape velocity— 30.13 miles/second

age— 6-10 million years

lifespan— less than 6 million 600 thousand years-11 million years

Main sequence (prime of its life) lifespan— less

than 6 years-10 million years

Time left as a main sequence star — has completed this phase of its life

When will this star die?— less than 600 thousand-1

million years Fate of the star— supernova, neutron star

A red supergiant with a bow shock is very rare. Betelgeuse and mu cephei are like this, but they are invisible. This is amongst the largest stars known.

*Star's name—
Betelgeuse
(alpha orionis)*

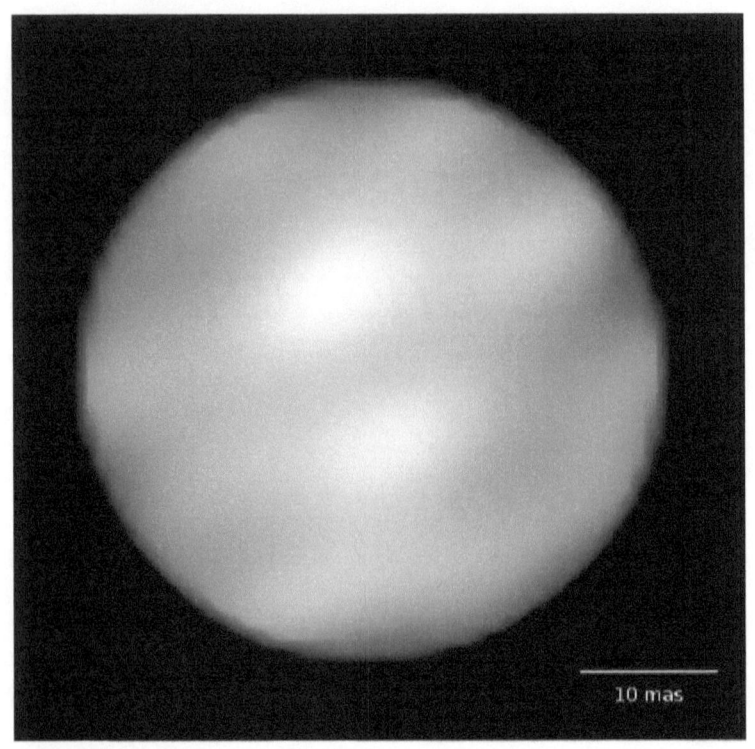

Stellar class—
M1-2(1a-1ab)

*location—
constellation of
Orion*

color— red

*Right ascension—
05 hours 55
minutes 10.30536
seconds*

declination— 07 degrees 24 minutes 25.4304 seconds

Apparent magnitude— .5

Absolute magnitude— -5.85

distance—723.72 light years

mass— 7.7-16.6 (11.6) suns

radius— 887 suns

luminosity— 90,000-150,000 suns

(18,716 suns)

Surface temperature— 3,590 degrees kelvin

Velocity through space— 13.5842 miles/ second recession

Movement in arcseconds/year— .0297 arcseconds/year

Direction on the unit circle— 19.97 degrees

parallax— 4.51 milliarcseconds

size— 3.995 arc seconds

gravity— .66-1.425 millimeters/second^2

Escape velocity— 25.26-37.08 miles/second

age— 8 million-8 million 500 thousand years

lifespan— less than 8 million 800 thousand years-9 million 350 thousand years (9 million 75 thousand years)

Main sequence (prime of its life) lifespan— less than 8 years-8 million 500 thousand years

Time left as a main sequence star — has completed this phase of its life (left the main sequence 1 million years ago when it was 8 million 75 thousand years old.)

When will this star die?— less than 907,500 years (less than a million years.)

it entered the supergiant phase

of its life, 40,000 years ago, when it was 9 million 115 thousand years old. from a red super giant star, it will be briefly a yellow super giant star, then explode as a blue super giant or a wolf rayed star. It could explode at any time and reach an apparent magnitude of -10 to -12.4. it will explode in 100,000-1,000,000 years.

Fate of the star—type 2p supernova, neutron star, with 2 x 10^46 joules of neutrinos and kinetic energy 2 x 10^44 joules could supernova any time, reaching apparent magnitude -10 to -12.4, .83-10.97 times fainter than a *full moon, 75.76-691 times brighter than venus, and 2,292-20,906 times brighter than Sirius.) It is the nearest star to us* which is a supernova candidate. it is so large that it would reach out past the orbit of jupiter.

Its rotation time is 8.4 years. It could contain 1.6 billion

to 2 trillion earths.
It is losing 1 solar
mass every 10,000
years.

Star's name—
antares a

*Stellar class—
M1.5(ab- ab)/
B2.5(5)*

*location—
constellation of
scorpius*

*color— red/blue
white green*

*Right ascension—
16 hours 29
minutes 24.45970
seconds*

declination— -25 degrees 25 minutes 55.2094 seconds

Apparent magnitude— .6 to 1.6/5.5

Absolute magnitude— -5.28/-.895

distance—550 light years

mass— 12 (11-14.3)/7.2 suns

radius— 680/5.2 suns

luminosity— 97,700 (10,000 in visible)/195 suns

Surface temperature— 3,570/18,500 degrees kelvin

Velocity through space— 2.108

miles/second approach

Movement in arcseconds/year— .026 arcseconds/year

Direction on the unit circle— 207.46 degrees

parallax— 5.89 milliarcseconds

size— 4.03 arc seconds

gravity— 7.03 millimeters/second^2

*Escape velocity—
36.01 miles/
second*

*age— 10 million
900 thousand
years-11 million
300 thousand
years*

lifespan— less than 11 million 990 thousand years-12 million 430 thousand years

Main sequence (prime of its life) lifespan— less than 10 million 900 thousand years-11 million 300 thousand years

Time left as a main sequence star — has completed this phase of its life

When will this star die?— less than 1

million 90 thousand years-1 million 130 thousand years

in 100,000 years, Antares may supernova. It will reach absolute magnitude of -16 to -18.67, apparent magnitude of -9.86 to -12.53, and luminosity of 216 million to 2.51 billion suns.

Fate of the star— supernova, neutron star

4th nearest supernova candidate star. Size reaches out to between mars and jupiter. Progenitor star 17 sun masses with an age of 12 million *years, or 15 suns and age 11-15 million years. Separation of binary stars 2.86 arc seconds, or 529 Au (49.197 billion miles) and* an orbital period of 12,167 years.

Star's name—
deneb

Alpha cygni

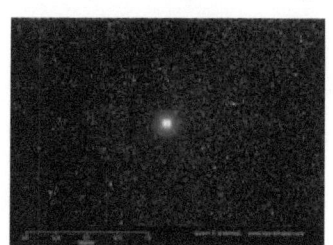

Stellar class—A2(1a)

location—constellation of cygnus

color— blue white

Right ascension— 20 hours 41 minutes 25.9 seconds

declination— 45 degrees 16 minutes 49 seconds

Apparent magnitude— 1.25

Absolute magnitude— -8.38

distance— 2,614.52 light years

mass— 19 suns (15-23 suns)

radius— 203 suns

luminosity— 190,000 suns (196,000 suns)

Surface temperature— 8,526 degrees kelvin

Velocity through space— 2.79 miles/second approach

Movement in arcseconds/year— .002786 secseconds/ year

Direction on the unit circle— 44.42 degrees

parallax— 2.29 milliarcseconds

size— .253 arc seconds

gravity— 4.92 inches/ second^2

Escape velocity— 82.93 miles/second

age— less than 5 million 720 thousand years

lifespan—less than 6 million 355 thousand years

Main sequence (prime of its life) lifespan— less than 5 million 720 thousand years

Time left as a main sequence star — has completed this phase of its life

When will this star
die?— less than
636 thousand
years

Fate of the star—
supernova,
neutron star

Deneb is one of
the largest white

stars. It loses 8+/-3 x 10^-7 solar masses/year, or a sun's mass every 100,000 years, and an earth mass every 500 years (the weight of 8,500 washington *monuments per second). Much of its early life was a type O star at mass 23 suns. It will eventually become an M1a star, and within a few million years, a supernova.*

Star's name — LBV 1806-20

Infrared image of Cluster 1806-20. LBV 1806-20 is the brightest star, on the left.

*Stellar class—
O9-B2*

*location—
constellation of
sagittarius*

color— blue

*Right ascension—
18 hours 08*

minutes 14.31 seconds

declination— -20 degrees 24 minutes 41.1 seconds

Apparent magnitude— 3.79

Absolute magnitude— -10.92

distance—40,000 light years

mass— 36 suns

radius— 200 suns

luminosity— 2 million suns

Surface temperature— 18,000-32,000 degrees kelvin

Velocity through space— no information found

Movement in arcseconds/year—

no information found

Direction on the unit circle— no information found

parallax— .115 milliarcseconds

size—.02299 arc seconds

gravity— 9.6 inches/second^2

Escape velocity— 115.01 miles/second

age— 3 million years-4 million 500 thousand years

lifespan—less than 3 million 300 thousand years-4 million 495 thousand years

Main sequence (prime of its life) lifespan— less than 3 million years-4 million 500 thousand years

Time left as a main sequence star — has completed this phase of its life

When will this star die?— less than 300 thousand

years, or at any time

Fate of the star—supernova, black hole

Star's name — epsilon aurigae almaaz

Stellar class — F0(1ab)/ B5(5)

*location—
constellation of
auriga*

color— white

*Right ascension—
05 hours 01
minutes 58.13245
seconds
declination— 43*

degrees 49 minutes 23.9059 seconds

Apparent magnitude— 2.98

Absolute magnitude— -9.1

distance— 2,129-4,890 light

years (2,000 light years)

mass— 15 (2.2-15)/6-14 suns

radius— 143-358/3.9 suns

luminosity— 373,483 suns

Surface temperature— 7,750/15,000 degrees kelvin

Velocity through space— 1.55 miles/second approach

Movement in arcseconds/year— .0028 arcseconds/year

Direction on the unit circle— 197.92 degrees

parallax— .115 milliarcseconds

size— .095-.548 arc seconds

gravity— 1.25-7.82 inches/ second^2

Escape velocity— 66.33 miles/ second

age— less than 11 million 480 thousand years

lifespan—less than 11 million 480 thousand years

Main sequence (prime of its life) lifespan— less

than 10 million 328 thousand years

Time left as a main sequence star — has completed this phase of its life

When will this star die?— less than 152,000 years

Fate of the star—supernova, neutron star

The orbital period of this binary star system is 9.896 days with a separation distance of 8.389 million miles (??)

Star's name — WR 102ka peony star

WR 102ka

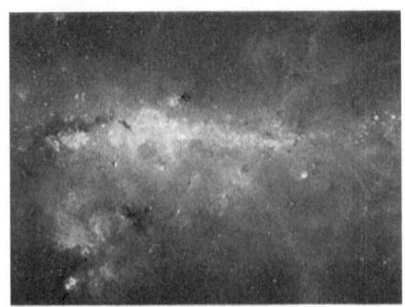

The "Peony Nebula," as discovered by NASA's Spitzer

Space Telescope. This three-color infrared composite shows 3.6-micrometre light in blue, 8-micrometre light in green, and 24-

micrometre light in red. The Peony nebula is the reddish cloud of dust in and around the white circle, surrounding the Peony nebular star.

Stellar class— OFpe/Wn9

location— constellation of sagittarius

color— blue

Right ascension— 17 hours 46 minutes 18.12 seconds

declination— -29 degrees 01 minutes 36.5 seconds

Apparent magnitude— 10.64

Absolute magnitude— -11.43

distance—26,000 light years

mass— 100 suns

radius— 92 suns
luminosity— 3.2 million suns
Surface temperature— 25,100 degrees kelvin

Velocity through space— no information found

Movement in arcseconds/year— no information found

Direction on the unit circle— no information found

parallax— .125 milliarcseconds

size— .011535 arc seconds

gravity— 10 feet 6 inches/ second^2

*Escape velocity—
282.62 miles/
second*

*age— less than 3
million years*

*lifespan—less than
3 million 300
thousand years*

Main sequence (prime of its life) lifespan— less than 3 million years

Time left as a main sequence star — has completed this phase of its life

When will this star die?— less than 300 thousand years

Fate of the star— supernova, black hole

This star is one of several candidates

for the most luminous star known. WR25 is even more luminous.

*Star's name—
polaris

alpha ursae
minoris*

Stellar class— F7(1b)/ F6(5)/F3(5)

location— constellation of ursa minor

color— white

Right ascension— 02 hours 31

minutes 49.09 seconds

declination— 89 degrees 15 minutes 50.8 seconds

Apparent magnitude— 1.98/9.2/8.7

Absolute magnitude— -3.63/3.6/3.1

distance—323-433 light years (433.8 light years)

mass— 5.4/1.26/1.39 suns

radius—37.5/1.04/1.38 suns

luminosity—2,356/3.11/4.92 suns

Surface temperature— 6,015/6,816/7,004 degrees kelvin

Velocity through space— 10.54 miles/ second approach

Movement in arcseconds/year— .046 arcseconds/year

Direction on the unit circle— 345.08 degrees

parallax— 7.54 milliarcseconds

size— .2823-.3785 arc seconds

gravity— 3 feet 4.95 inches/second^2

Escape velocity— 102.87 miles/second

age— 6.8 billion years (all 3 stars) lifespan— over 6.8 billion years

Main sequence (prime of its life) lifespan— less than 69 million 300 thousand years

Time left as a main sequence star — has completed this phase of its life

When will this star die? — less than 74.8 million years

*Fate of the star—
planetary nebula,
white dwarf*

*This is the north
pole star.*

*Star's name—
alnilam

epsilon orionis*

*Stellar class—
B0(1a)*

*location—
constellation of
orion*

color— blue

*Right ascension—
05 hours 36*

minutes 12.8 seconds

declination— -01 degrees 12 minutes 06.9 seconds

Apparent magnitude— 1.7

Absolute magnitude— -6.89
distance—1,344 light years
mass— 40 suns (30-64.5 suns)
radius— 32.4 suns
luminosity— 537,000 suns

(275,000-832,000 suns)

Surface temperature— 27,500 degrees kelvin

Velocity through space— 16.06

miles/ second recession

Movement in arcseconds/year— .00183 arcseconds/year

Direction on the unit circle— 324.57 degrees

parallax— 1.65 milliarcseconds

size— .0528 arc seconds

gravity— 33 feet 10.34 inches/second^2

Escape velocity— 301.2 miles/ second

age— 5 million 700 thousand years

lifespan—less than 6 million 270 thousand

years

Main sequence (prime of its life) lifespan— less than 5 million 700 thousand years

Time left as a main sequence star — has completed this phase of its life

When will this star die?— less than 570 thousand years

Fate of the star— supernova, black hole

Star's name— rigel beta orionis

Stellar class— B8(1a)/ B9(5)/ B9(5)

*location—
constellation of
orion*

color— blue

*Right ascension—
05 hours 14
minutes 32.2721
seconds*

declination— -08 degrees 12 minutes 05.8981 seconds

Apparent magnitude— .13/6.67

Absolute magnitude— -7.84/-2.99

distance—860 light years

mass— 23 (18)/ 3.84/2.94 suns

radius— 78.9/ approximately 6 for other 2 stars suns

luminosity— 117,017 (120,000-279,000)/ 1,348.5 suns

Surface temperature — 12,100/14,500 degrees kelvin

Velocity through space — 12 miles/second recession

Movement in arcseconds/year—.0014 arcseconds/year

Direction on the unit circle—20.89 degrees

parallax— 3.78

milliarcseconds

size— .3 arc seconds

gravity— 3.28 feet/second^2

Escape velocity— 146.36 miles/second

age— 7-9 million years

lifespan—less than 7 million 700 thousand years-9 million 900 thousand years

Main sequence (prime of its life)

lifespan— less than 7 million years-9 million years

Time left as a main sequence star — has completed this phase of its life

When will this star die?— less than 700 thousand years, or any time

Fate of the star— type 2 supernova, black hole

Rigel is an alpha cygni variable star.

It is one of the closest possible supernova candidate stars. Stellar winds reach velocity of 186.41 miles/ second.

The periods of these stars are:

9.86 days, 18,000 years, 63 years, and separation distances of: 8.57 million, 59.836 billion, 1.43 billion miles

Star's name— alnitak zeta orionis

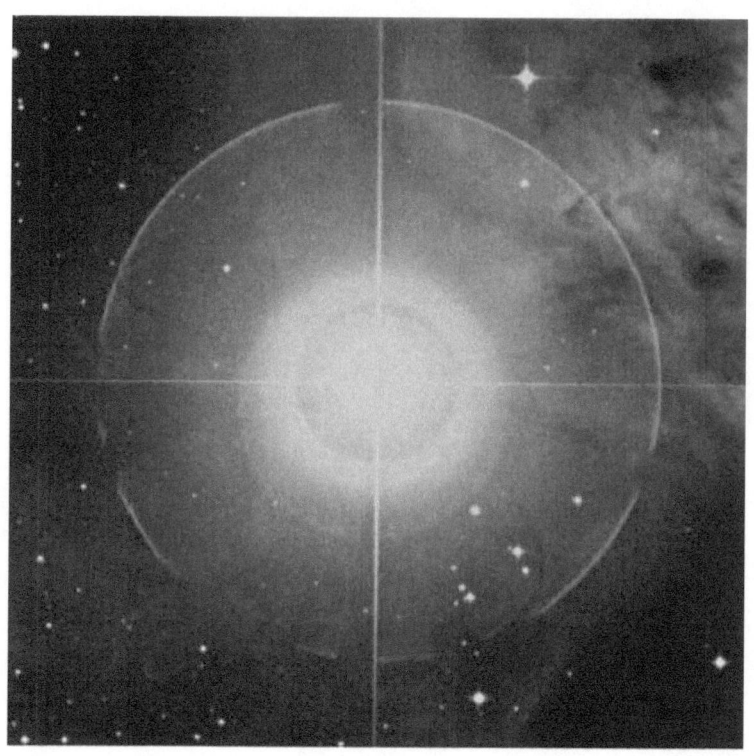

*Stellar class—
O9.5(1ab)/ B1(4)/ B0(3)*

*location—
constellation of orion*

color— blue

*Right ascension—
05 hours 40*

minutes 40.52666 seconds

declination— -01degrees 56 minutes 34.2649 seconds

Apparent magnitude— 1.77/4.04/3.84

Absolute magnitude— -6.0/-3.9/-4.1

distance—1,260 light years

mass— 33 suns

radius— 20 suns

luminosity— 250,000/3,106/3,734 suns

Surface temperature— 29,500/29,000 degrees kelvin

Velocity through space— 11.47

miles/ second recession

Movement in arcseconds/year—.00378 arcseconds/year

Direction on the unit circle—32.47 degrees

parallax— 5.02 milliarcseconds

size—

.05175 arc seconds

gravity— 77.316 feet/ second^2

Escape velocity—348.21 miles/second

age— 6 million 400 thousand/7 million 200 thousand/7 million years

lifespan—less than 7 million 40

thousand/7 million 920 thousand/7 million 700 thousand years

Main sequence (prime of its life) lifespan— less than 6 million 336 thousand/7 million

200 thousand/7 million years

Time left as a main sequence star — has completed this phase of its life

When will this star die?— less than 640 thousand/ 720

thousand/ 700 thousand years

Fate of the star— supernova, black hole

Star's name— saiph kappa orionis

Stellar class— B8(1a)

*location—
constellation of
orion*

color— blue

*Right ascension—
05 hours 47
minutes
45.36884 seconds*

declination— -09 degrees 40 minutes 10.5777 seconds

Apparent magnitude— 2.09

Absolute magnitude— -6.1

distance—650 light years

mass— 15.5 suns

radius— 22.2 suns

luminosity— 56,881 suns

Surface temperature— 26,500 degrees kelvin

Velocity through space— 12.4 miles/second recession

Movement in arcseconds/year— .00194 arcseconds/year

Direction on the unit circle— 318.76 degrees

parallax— 5.02 milliarcseconds

size— .111 arc seconds

gravity— 27 feet 11.39 inches/second^2

Escape velocity— 226.51 miles/second

age— 10 million 600 thousand-11 million 600 thousand years

lifespan—less than 11 million 660 thousand years-12 million 740 thousand years

Main sequence (prime of its life) lifespan— less than 10 million 494 thousand years-11 million 484 thousand years

Time left as a main sequence star —

has completed this phase of its life

When will this star die?— less than 1 million 60 thousand years-60 thousand years

Fate of the star—supernova, neutron star

Mass lose from stellar winds 9×10^{-7} solar masses/year, and is equal to 1 solar mass every 1.1 *million years. (the weight of 1.4 million washington monuments per second.)*

*Star's name—
V838 monocerotis*

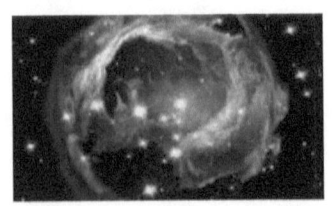

Stellar class— M6.3L/ B3(5)

location— constellation of monoceros

color— red

Right ascension— 07 hours 04

minutes 04.85 seconds

declination— -03 hours 50 minutes 50.1 seconds

Apparent magnitude— 6.75 to 15.6

Absolute magnitude— -5.61 (-9.8 maximum)

distance—19,886 light years

mass— estimate 25.77/5-10 suns

radius— 380 (1,570 maximum) suns

luminosity— 15,000 (1million)suns

Surface temperature—

3,270 degrees kelvin

Velocity through space— no information found

Movement in arcseconds/year— no information found

Direction on the unit circle— no information found

parallax—.164 milliarcseconds

size— .257323 arc seconds

gravity— 2.83 millimeters/ second^2

Escape velocity— 34.73 miles/ second

age— 4 million years

lifespan— less than 4.4 million years

Main sequence (prime of its life) lifespan— less than 4 million years

Time left as a main sequence star — has completed this phase of its life

When will this star die? — less than 400 thousand years

Fate of the star—supernova, black hole

One of the most luminous stars known. Opaque ejected dust completely engulfed b type companion star.

www.ingramcontent.com/pod-product-compliance
Lightning Source LLC
Chambersburg PA
CBHW030942240526
45463CB00016B/1155